I0470789

Special Forces
Tracking and Countertracking

September 2009

Headquarters, Department of the Army

TC 31-34-4

SPECIAL FORCES
TRACKING & COUNTERTRACKING

TC 31-34-4
SPECIAL FORCES
TRACKING AND COUNTERTRACKING

This Edition Copyright © 2013 by Special Operations Press

Cover, Layout and Design by: Special Operations Press

ISBN-13: 978-1481837736

ISBN-10: 1481837737

Proudly Printed in the
U.S.A

Training Circular
31-34-4

Headquarters
Department of the Army
Washington, DC, 30 September 2009

Special Forces
Tracking and Countertracking

Contents

Figures

Tables

Preface

This training circular (TC) provides a doctrinal framework for Special Forces (SF) personnel involved in tracking and countertracking operations.

PURPOSE

Tracking, countertracking, and dog-tracker team operations are basic and fundamental to every SF operation whether offensive or defensive in nature. This TC describes and illustrates how to track, how to avoid being tracked, and the theory behind the use of dog-tracker teams. Appendixes A and B provide SF Soldiers with sample tracking logs for their use. This TC does not describe specific electronic-tracking techniques, such as transistor-transistor logic, cell phone triangulation, or other sophisticated electronic-tracking tactics, techniques, and procedures (TTP), but it does introduce current doctrine that addresses those topics. This TC provides the basis for common SF tactical application primarily in a rural environment and it briefly discusses urban tracking using dog teams.

SCOPE

SF Soldiers routinely employ unconventional tactics and techniques while conducting operations unilaterally and with indigenous assistance. The conduct of SF differs from conventional operations in the degree of political risk, operational techniques, independence from friendly support, and dependence on detailed operational intelligence and indigenous assets. The success of SF operations within these parameters depends greatly on the team's ability to read "signs" for defensive purposes and its ability to use that same knowledge to minimize the signs it leaves when transitioning through hostile terrain.

APPLICABILITY

This publication applies to the Active Army, the Army National Guard (ARNG)/Army National Guard of the United States (ARNGUS), and the United States Army Reserve (USAR) unless otherwise stated.

ADMINISTRATIVE INFORMATION

Unless this publication states otherwise, masculine nouns and pronouns do not refer exclusively to men. The proponent of this manual is the United States Army John F. Kennedy Special Warfare Center and School (USAJFKSWCS). Submit comments and recommended changes on DA Form 2028 (Recommended Changes to Publications and Blank Forms) and send directly to Commander, USAJFKSWCS, ATTN: AOJK-DTD-SF, Fort Bragg, NC 28310-9610.

Chapter 1

Tracking

INTRODUCTION

1-1. This TC describes traditional tracking through rural terrain and, to a lesser extent, discusses urban tracking. Understanding the ancient art and science of tracking lays a firm foundation for a variety of activities useful in SF operations. A Soldier trained in tracking techniques can use deception maneuvers that minimize telltale signs and throw off or confuse poorly trained trackers who do not have the experience to spot the signs of a deception.

1-2. An understanding of the thought processes and the TTP used in traditional tracking are useful as an adjunct to—and a basis for—understanding an array of modern, technologically-based tracking activities. Traditionally, tracking has been defined as the art of being able to follow people or animals by the signs they leave when they move. Today, it is possible to track the enemies of the United States through electronic means, such as the equipment used with sensitive site exploitation (SSE) techniques, as well as tagging, tracking, and locating TTP. Not all enemies the United States encounters ever set foot on a traditional battlefield, so it is important to remember, no matter how they choose to negotiate or communicate from one location to another, it is nearly impossible to do so without leaving signs behind.

1-3. Trained and experienced trackers can detect the signs left behind, no matter how small. This is especially true in urban environments through forensic and biometric means, including the electronic tracking of e-mail and financial transactions, both of which leave behind electronic signs called *cookies* or *breadcrumbs*. Soldiers can find additional information regarding other forms of technological tracking in current SSE lessons learned, and other SF doctrinal publications.

1-4. A successful tracker must—

- Be patient and consistent.
- Move slowly, quietly, and steadily while simultaneously detecting and interpreting signs.
- Avoid fast movement that may cause him to overlook or lose signs, or to walk into an enemy ambush.
- Be persistent and be able to continue the mission when signs are lost or scarce because of bad weather or terrain.
- Be observant and be able to see things not obvious at first glance.
- Use his sense of smell and hearing to augment his sight and intuition.
- Develop his intuition and a feel for things that do not *look right*. This ability may help him regain a lost trail or discover additional signs.

1-5. As a tracker follows a trail, he uses the above-mentioned skills to build a picture of the enemy in his mind while asking himself these questions:

- How many people am I following?
- Are they male or female?
- Are they adults or children?
- What is their state of training?
- How are they equipped?
- Are they healthy?
- What is their state of morale?
- Do they know they are being followed?
- Are they familiar with the area?

1-6.   To answer these questions, the tracker uses available indicators (Figure 1-1), such as signs that tell an action occurred at a specific time and place. By comparing indicators, the tracker obtains answers to his questions.

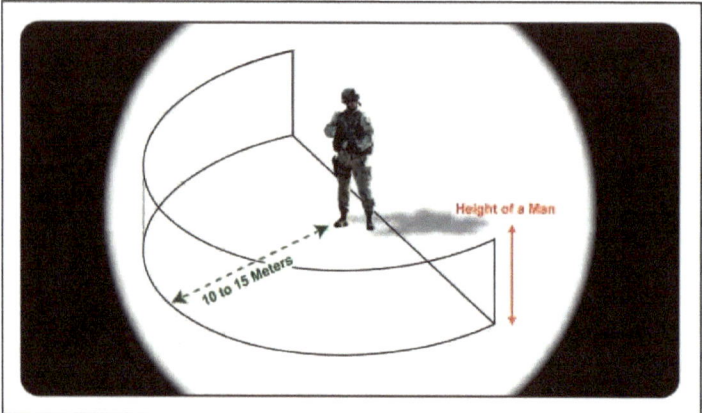

Figure 1-1. Area a tracker surveys to find tracking indicators

TRACKING SIGNS

1-7.   Signs are visible marks left by individuals or animals as they pass through an area. The tracker must know the following categories of signs:

- *Ground Signs.* These are signs left below the knees. All ground signs are further divided as follows:
  - *Large Signs.* These are signs caused by the movement of 10 or more individuals through the area.
  - *Small Signs.* These are signs caused by the movement of one to nine individuals through the area.
- *High Signs (also known as top signs).* These are signs left above the knees. They are also divided into large and small signs.
- *Temporary Signs.* These signs will eventually fade with time (for example, a footprint).
- *Permanent Signs.* These signs require weeks to fade or will leave a mark forever (for example, broken branches or chipped bark).

TRACKING INDICATORS

1-8.   One of six tracking indicators defines any signs the tracker discovers. Those indicators are: displacement, stains, weathering, odor, litter, and immediate-use intelligence.

DISPLACEMENT

1-9.   Displacement occurs when anything is moved from its original position. A good example of displacement is a well-defined footprint in soft, moist ground (Figure 1-2, page 1-3). The footgear or bare feet of the person who left the print displaced the soil by compression, leaving an indentation in the ground. The tracker can study this sign and determine several important facts. For example, a print left by worn footgear or by bare feet may indicate lack of proper equipment.

1-10.   Displacement can also results from clearing a trail by breaking or cutting through heavy vegetation with a machete; these trails are obvious to the most inexperienced tracker. Individuals may unconsciously break more branches as they move behind someone who is cutting a path.

Figure 1-2. Footprint in soft, moist ground conditions

1-11. Persons carrying heavy loads who stop to rest can also make displacement indicators. Prints made by box edges can help to identify the load. When loads are set down at a rest halt or campsite, they usually crush grass and twigs (Figure 1-3). A reclining man can also flatten the vegetation (Figure 1-4, page 1-4).

Figure 1-3. Print made by "box edge," indicating equipment (rifle stock)

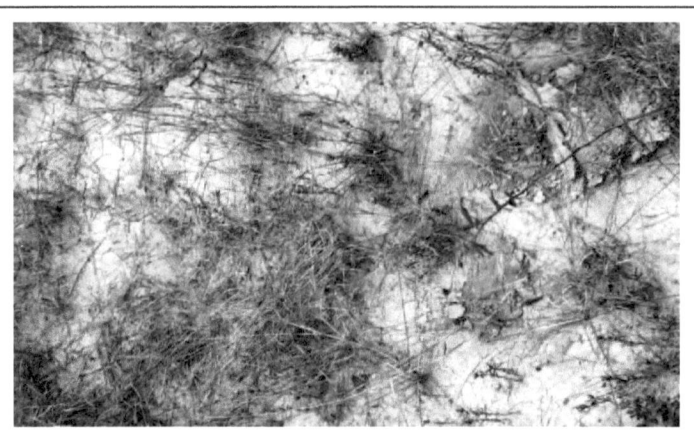

**Figure 1-4. Crushed vegetation with toe scuffs, indicate a person had taken
a prone posture (feet to the upper right, head to the lower left)**

Analyzing Footprints

1-12. Footprints can indicate the direction, rate of movement, number, sex, and whether the individual knows he is being tracked. The footprint can be a whole print but is usually only the "heel dig" and "toe push" footprint. Trackers can find footprints on the underside of large leaves that have not dried out and are lying on the ground.

1-13. If footprints are deep and the pace is long, rapid movement is apparent. Extremely long strides and deep prints, with toe prints deeper than heel prints, indicate running (Figure 1-5). Prints that are deep with a short, narrowly paced stride and show signs of shuffling indicate the person who left the print is carrying a heavy load (Figure 1-6, page 1-5).

Figure 1-5. Footprints left by someone running

Figure 1-6. Footprints left by someone carrying a heavy load

1-14. If the party members realize they are being followed, they may try to hide their tracks. Persons walking backward have a short, irregular stride (Figures 1-7 through 1-9, pages 1-5 and 1-6). The prints have an unnaturally deep toe, and soil is displaced in the direction of movement. These types of prints are characterized by "toe digs" and "heel push," as opposed to the normal footprint.

Figure 1-7. Footprints left by someone walking backward in dry sand

Figure 1-8. Footprints left by someone walking backward in wet sand

Figure 1-9. Footprints left by someone walking backward over vegetation

1-15. To determine the sex of a member of the party being followed the tracker should study the size and position of the footprints (Figure 1-10, page 1-7). Women tend to be pigeon-toed; men walk with their feet straight ahead or pointed slightly to the outside. Prints left by women are usually smaller and the stride is usually shorter than that taken by men.

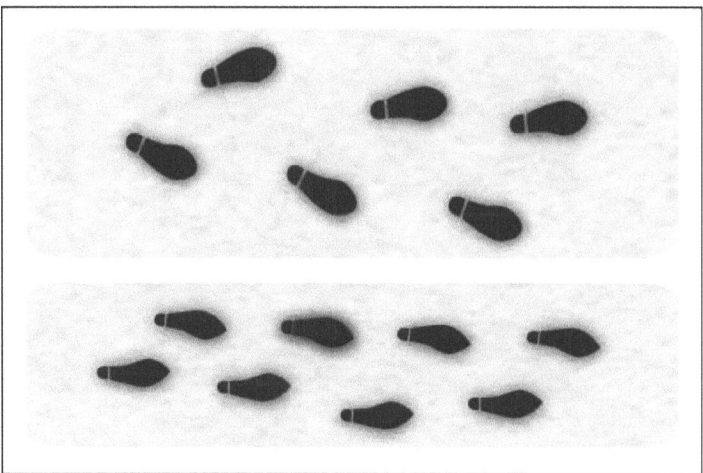

Figure 1-10. Man versus woman

Determining Key Prints

1-16. Normally, the last man in the file leaves the clearest footprints; these should be the key prints. The tracker can use several techniques to create a measuring/tracking stick to aid him in locating these tracks. One way is for the tracker to cut a stick to match the length of the prints and notch it to show the length of the strides (generally 36 inches) and the length and widest part of the sole of the key prints. The tracker then studies the angle of the key prints in relation to the direction of movement. To identify key prints, the tracker looks for an identifying mark or feature, such as worn or frayed footgear. If the trail becomes vague, erased, or merges with another, the tracker can use his stick-measuring device and identify the key prints after close study. Another technique is for the tracker to use his weapon/rifle as a measuring device; this allows him to engage threats more rapidly while tracking hostile forces. This method helps the tracker stay on the trail. By using the box method, he can count up to 18 persons. The tracker can also—

- Use the stride as a unit of measure when determining key prints (Figure 1-11, page 1-8). The tracker uses these prints and the edges of the road or trail to box in an area to analyze.
- Use the 36-inch box method if key prints are not evident (Figure 1-12, page 1-8). To use this method, the tracker uses the edges of the road or trail as the sides of the box. He measures a cross section of the area 36 inches long, counting each indentation in the box and dividing by two. This method gives a close estimate of the number of individuals who made the prints; however, it is not as accurate as the stride measurement.

Recognizing Other Signs of Displacement

1-17. Foliage, moss, vines, sticks, or rocks that are scuffed or snapped from their original position form valuable indicators. Broken dirt seals around rocks, mud or dirt moved to rocks or other natural debris, and water moved onto the banks of a stream are also good indicators (Figures 1-13 through 1-17, pages 1-9 through 1-11). Vines may be dragged, dew droplets displaced, or stones and sticks overturned to show a different color underneath. Grass or other vegetation may be bent or broken in the direction of movement.

1-18. The tracker inspects all areas for bits of torn clothing, threads, or dirt from footgear, which can fall and be left on thorns, snags, or on the ground. Flushed from their natural habitat, wild animals and birds are also examples of displacement. Cries of birds excited by unnatural movement are an indicator; moving tops of tall grass or brush on a windless day indicate that something is moving the vegetation.

1-19. Changes in the normal life of insects and spiders may indicate someone has recently passed. Valuable clues are disturbed bees, ant holes covered by someone moving over them, or torn spider webs. Spiders often spin webs across open areas, trails, or roads to trap flying insects. If the tracked person does not avoid these webs, he will leave an indicator behind.

1-20. If the person being followed tries to use a stream to cover his trail, the tracker can still follow successfully. When a person loses his footing or walks carelessly, he can displace algae and other water plants. He can also displace or overturn rocks from their original position to indicate a lighter or darker color on the opposite side. The person entering or exiting a stream creates slide marks or footprints, or scuffs the bark on roots or sticks (Figures 1-18 through 1-22, pages 1-11 through 1-13). Normally, a person or animal seeks the path of least resistance; therefore, when searching the stream for an indication of departures, trackers will find signs in open areas along the banks.

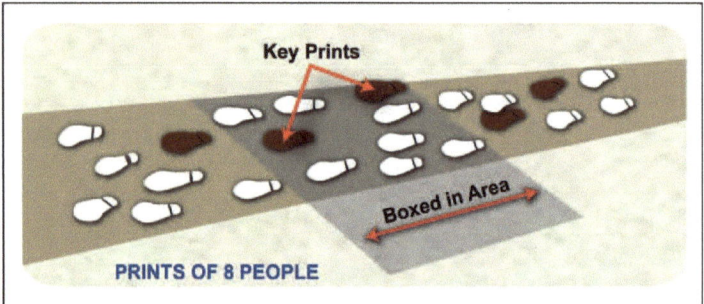

Figure 1-11. Using the stride as a unit of measure

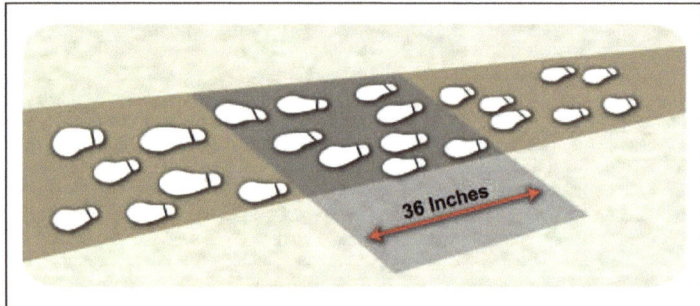

Figure 1-12. Using the 36-inch box method

Figure 1-13. Broken dirt seal around rocks indicating disturbance

Figure 1-14. Turned over rocks and sticks

Figure 1-15. Crushed or disturbed vegetation on a creek bank

Figure 1-16. Crushed or disturbed vegetation "high sign"

Figure 1-17. Crushed or disturbed vegetation

Figure 1-18. Disturbance at water crossing approximately 15 minutes
after the fact

Figure 1-19. Staining and displacement in water

Figure 1-20. Slip marks and water-filled footprints on stream bank
indicating movement out of the creek

Figure 1-21. Slip marks and water-filled footprints on stream bank
looking from the bank into the creek

Figure 1-22. Slip marks and water-filled footprints on stream bank
looking from the creek onto the bank

STAINS

1-21. A stain occurs when any substance from one organism or article is smeared or deposited on something else. The best example of staining is blood from a profusely bleeding wound. Bloodstains often appear as spatters or drops and are not always on the ground; they also appear smeared on leaves or twigs of trees and bushes. The tracker can determine the seriousness of the wound and how far the wounded person can move unassisted. This process may lead the tracker to enemy bodies or indicate where they have been carried.

1-22. By studying bloodstains, the tracker can determine the wound's location, as follows:
- If the blood seems to be dripping steadily, it probably came from a wound on the trunk.
- If the blood appears as if it was slung toward the front, rear, or sides, the wound is probably in the extremity.
- Arterial wounds appear to pour blood at regular intervals as if poured from a pitcher. If the wound is venous, the blood pours steadily.
- A lung wound deposits pink, bubbly, and frothy bloodstains.
- A bloodstain from a head wound appears heavy, wet, and slimy.
- Abdominal wounds often mix blood with digestive juices so the deposit has an odor and is light in color.

1-23. Any body fluids (such as urine, blood, or feces) deposited on the ground, trees, bushes, or rocks will leave stains. A field-expedient technique for determining a blood trail is to spray hydrogen peroxide on the suspected stain. If the hydrogen peroxide bubbles/foams rapidly it is an indicator of blood. Trackers can test this technique in specific environmental and weather conditions during rehearsals and proficiency training.

1-24. On a calm, clear day, the leaves from bushes and small trees turn so their dark top sides show. However, when a man passes through an area and disturbs the leaves, he will generally cause the lighter side of the leaf to show. This movement is also true with some varieties of grass. Moving causes an unnatural discoloration of the area, which is called "shine" (Figure 1-23). Grass or leaves that have been stepped on will have a bruise on their lighter sides.

Figure 1-23. Example of shine

1-25. Staining can also occur when a person drags muddy footgear over grass, stones, and shrubs. Thus, staining and displacement combine to form a movement and direction. Crushed leaves may stain rocky ground that is too hard to show footprints. Roots, stones, and vines may be stained where leaves or berries are crushed by moving feet (Figures 1-24 through 1-26, pages 1-15 and 1-16).

Figure 1-24. Staining on shoulder of dirt roadway (right to left)

Figure 1-25. Staining on roadway (left to right)

Figure 1-26. Staining caused by differing soil types

1-26. The tracker may have difficulty determining the difference between staining and displacement since both terms can be applied to some indicators. For example, muddied water may indicate recent movement; displaced mud also stains the water. Muddy footgear can stain stones in streams, and algae can be displaced from stones in streams and can stain other stones or the bank (Figure 1-27, page 1-17). Muddy water collects in new footprints in swampy ground; however, the mud settles and the water clears with time. The tracker can use this information to indicate time. Normally, the mud clears in about one hour, although time varies with the terrain. Since muddied water travels with the current, it is usually best to move downstream.

WEATHERING

1-27. Weathering either aids or hinders the tracker. It also affects indicators in certain ways so that the tracker can determine their relative ages. However, wind, snow, rain, or sunlight can erase indicators entirely and hinder the tracker. The tracker should know how weathering affects soil, vegetation, and other indicators in his area. He cannot properly determine the age of indicators until he understands the effects weathering has on trail signs.

1-28. For example, when bloodstains are fresh, they are bright red. Air and sunlight first change blood to a deep, ruby-red color, then to a dark brown crust when the moisture evaporates. Scuff marks on trees or bushes darken with time. Sap oozes on trees and then hardens when it makes contact with the air.

1-29. Weather greatly affects footprints (Figures 1-28 through 1-30, pages 1-18 and 1-19). Thus, by carefully studying the weathering process, the tracker can estimate the age of the footprints.

1-30. If particles of soil are just beginning to fall into the print, it is very recent. At this point, the tracker should focus on becoming a stalker. If the edges of the print are dried and crusty, the prints are probably about an hour old. This process varies with terrain and is only a guide.

1-31. A light rain may round the edges of the prints. By remembering when the last rain occurred, the tracker can place the prints into time frames. Heavy rains may erase all signs.

**Figure 1-27. Staining at creek
(center rock is wet while the remainder of the rocks are dry)**

1-32.   Trails exiting streams may appear weathered by rain due to water running from clothing or equipment running over the tracks. This trait is especially true if the party exits the stream single file. Then, each person deposits water into the tracks. The existence of a wet, weathered trail slowly fading into a dry trail indicates the trail is fresh.

1-33.   Wind dries out tracks and blows litter, sticks, or leaves into prints. By recalling wind activity, the tracker can estimate the age of the tracks. For example, the tracker may reason, "the wind is calm at present but blew hard about an hour ago. These tracks have litter blown into them, so they must be over an hour old." However, he must be sure the litter was blown into the prints and was not crushed when they were made or thrown on the track in an effort to conceal them.

1-34.   Wind affects sound and odors. If the wind is blowing down the trail (toward the tracker) sounds and odors may be carried to him. Conversely, if the wind is blowing up the trail (away from the tracker) he must be extremely cautious since wind also carries sound toward the enemy. The tracker can determine wind direction by dropping a handful of dust or dried grass from shoulder height. By pointing in the same direction the wind is blowing, the tracker can localize sounds by cupping his hands behind his ears and turning slowly. When the sounds are the loudest, the tracker is facing the origin.

1-35.   In calm weather (no wind), air currents that may be too light to detect can carry sounds to the tracker. Air cools in the evening and moves downhill toward the valleys. If the tracker is moving uphill late in the day or night, air currents will probably be moving toward him if no other wind is blowing. As the morning sun warms the air in the valleys, it moves uphill. The tracker considers these factors when plotting patrol routes or other operations. If he keeps the wind in his face, sounds and odors will be carried to him from his objective or from the party being tracked.

1-36.   The tracker should also consider the sun. It is difficult to fire directly into the sun, but if the tracker has the sun at his back and the wind in his face, he has a slight advantage.

Figure 1-28. Weathered footprint

Figure 1-29. Fresh footprint

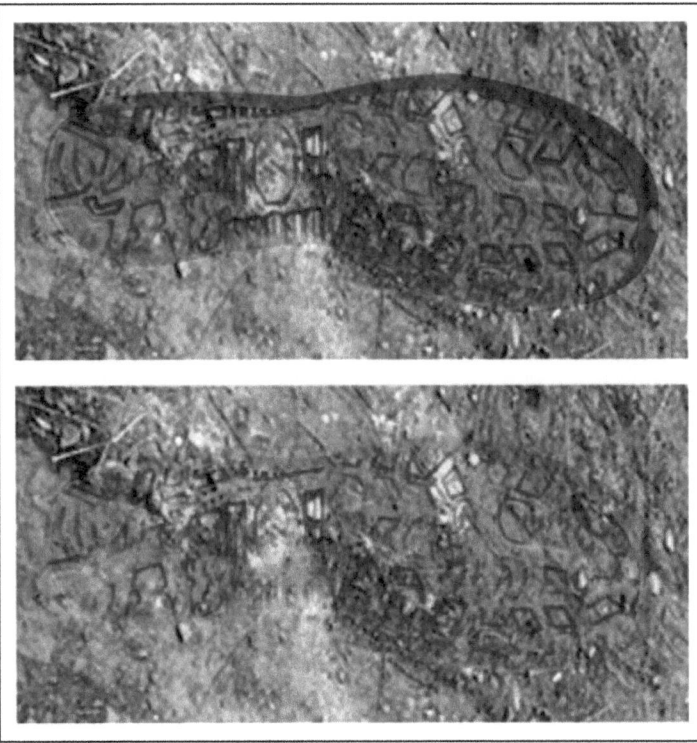

Figure 1-30. Effects of weather on the clarity of footprints

ODORS

1-37. The tracker should consider dietary habits, such as—
- Red meats.
- Tobacco.
- Alcohol.
- Certain spices (for example, those used in Korea).

1-38. With the convenience of an Army and Air Force Exchange Service facility while deployed, many Soldiers purchase and use fragrant soaps and body washes. Soaps and body washes cause strong, peculiar odors that are excreted through the sweat glands. These scents will be particularly evident to dog teams, but can be strong enough for a person to detect.

LITTER

1-39. Litter consists of anything not indigenous to the area that is left on the ground. A poorly trained or poorly disciplined unit moving over terrain is apt to leave a trail of litter. Unmistakable signs of recent movement include gum or candy wrappers, ration cans or wrappers, cigarette butts, and the remains of fires, urine, and bloody bandages. Rain flattens or washes litter away and turns paper into pulp. Exposure to weather can cause ration cans to rust at the opened edge; then the rust moves toward the center. The tracker must consider weather conditions when estimating the age of litter. He can use the last rain or strong wind as the basis for determining a time frame.

1-40. The tracker should also know what wildlife is in the area. Even sumps, regardless of how well camouflaged, are a potential source of litter. The best policy to follow is for the Soldier to take everything he brings with him.

IMMEDIATE-USE INTELLIGENCE

1-41. The tracker combines all the indicators and interprets what he has seen to form a composite picture for on-the-spot intelligence. For example, indicators may show contact is imminent and require extreme stealth.

1-42. The tracker avoids reporting his interpretations as facts. He reports what he has seen, rather than stating that these things exist. There are many ways a tracker can interpret the sex and size of the party, the load, and the type of equipment. Time frames can be determined by weathering effects on indicators.

1-43. Immediate-use intelligence is information about the enemy that can be used to gain surprise, to keep him off balance, or to keep him from escaping the area entirely. The commander may have many sources of intelligence, such as reports, documents, or prisoners of war. These sources can be combined to form indicators of the enemy's last location, future plans, and destination.

1-44. Tracking gives the commander definite information on which to act immediately. For example, a unit may report there are no men of military age in a village. This knowledge is valuable, only if combined with other information to create a composite enemy picture in the area. Therefore, a tracker who interprets trail signs and reports he is 30 minutes behind a known enemy unit, moving north, and located at a specific location, gives the commander information on which he can act.

Chapter 2

Countertracking

INTRODUCTION

2-1.   There are two types of human trackers—combat trackers and professional trackers. Combat trackers look ahead for signs and do not necessarily look for each individual sign. Commanders use them for situations when speed is vital and the threat level is high. Professional trackers go from sign to sign. If they cannot find signs, they will stop and search till they find one.

2-2.   If an enemy tracker finds the tracks of two men, he may determine that a highly-trained specialty team is operating in his area. However, knowledge of countertracking enables the team to survive by remaining undetected.

2-3.   To confuse the combat tracker and throw him off track, the team always starts its movement away from its objective. The team uses the same maneuver when evading dogs. (Chapter 3 provides additional information on dogs.) The team should travel in a straight line for about an hour and then change direction. Changing course will cause the tracker to cast in different directions to find the track.

2-4.   Use of weather and times of day to defeat a dog-tracker team is also important. Weather tends to disintegrate tracks. Wind blows scents away, removing gases and particles that identify the track to the dog. Wind also aerates, dries out, and reduces bacterial activity. Although a low to moderate wind condition is an asset for air-scent search, the less wind the better is the rule for tracking.

2-5.   High humidity favors tracking. Humid conditions combined with high temperatures encourage discrimination in the tracking dog for a significant time after the track is laid. A light rain refreshes a track scent; however, pouring rain or heavy snow disintegrates a track. The sun can literally *burn out* a track. Direct sunlight kills bacteria and generates locally high temperatures that dry out and destroy scent producing reactions. The best tracking conditions usually are found in the early morning or evening, and at night (at night however, the handler is slower due to limited visibility) on north-facing slopes (Northern Hemisphere specific), across low areas, and in damp grasses.

CAMOUFLAGE

2-6.   Camouflage applies to tracking when the party being followed uses techniques to confuse or slow a tracker; for example, walking backward to leave confusing prints, brushing out trails, and moving over rocky ground or through streams. Camouflaged movement indicates a trained adversary (Figures 2-1 through 2-8, pages 2-2 through 2-5).

2-7.   The team being followed may use two types of routes to cover its movement. The team must also remember, that travel time increases when trying to camouflage the signs left during movement. Two types of routes include:

- *Most-Used Routes.* Movement on lightly traveled sandy or soft trails is easily tracked. However, a person may try to confuse the tracker by moving on hard-surfaced, often-traveled roads or by merging with civilians. The tracker should carefully examine these routes. If a well-defined approach leads to the enemy, it will probably be mined, ambushed, or covered by some form of security force.
- *Least-Used Routes.* These routes avoid all man-made trails or roads and are less confusing for the tracker. They are normally magnetic azimuths between two points.

EVASION

2-8. Evasion of the tracker or a pursuing element is a difficult task requiring the use of immediate-action drills designed to counter the threat. A team skilled in tracking techniques can successfully use deception drills to minimize signs left behind which the enemy can use against them. However, it is very difficult for a person, let alone a group, to move across any area without leaving signs noticeable to the trained eye.

Figure 2-1. Countertracking foot wrapping using a cravat
and a terrycloth towel

Figure 2-2. Countertracking foot wrapping using a cravat

Figure 2-3. Example of a footprint made by a boot
wrapped in a terrycloth towel

Figure 2-4. Difference between wrapped boot and the same person
without wrapping the boot (upper wrapped, lower unwrapped)

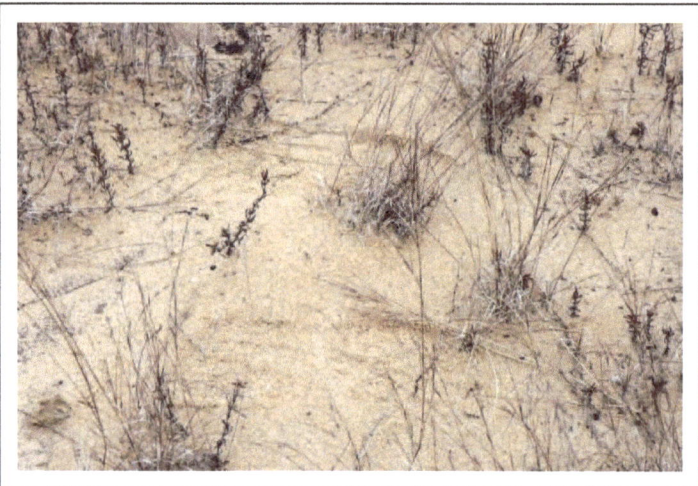

Figure 2-5. Examples of boot prints using towels as countertracking wraps (movement is from the left to the right as indicated by the pronounced edge)

Figure 2-6. Brushing out ground sign

Figure 2-7. Brushed-out ground sign

Figure 2-8. Evidence of countertracking attempts

REDUCTION OF TRAIL SIGNS

2-9. A team trying to hide its trail moves at reduced speed; therefore, enabling the experienced tracker to gain time. A team should use the following methods to reduce trail signs:

- Wrap footgear with rags or wear soft-soled sneakers that make footprints rounded and less distinctive.
- Change into footgear with a different tread immediately following a deceptive maneuver, which creates a gap or void in the noticeable track.
- Walk lightly on hard or rocky ground to reduce the possibility of displacing rocks or leaving scuffs for the tracker to follow.

DECEPTION TECHNIQUES

2-10. Evading a skilled and persistent enemy tracker requires the use of skillfully executed maneuvers to deceive the tracker and cause him to lose the trail. An enemy tracker cannot be outrun by a team carrying equipment because the tracker travels light and is escorted by enemy forces designed for pursuit. The size of the pursuing force dictates the team's chances of success in using ambush-type maneuvers. Teams can use some of the following techniques in immediate-action and deception drills.

Backward Walking

2-11. One of the most basic deception techniques is walking backward (Figure 2-9) in tracks already made and then stepping off the trail onto terrain or objects that leave little to no signs. Skillful use of this maneuver causes the tracker to look in the wrong direction once he has lost the trail. This maneuver must be used in conjunction with another deception technique, such as those described in paragraph 2-9. However, the maneuver will probably fail if a professional tracker is following the team's trail.

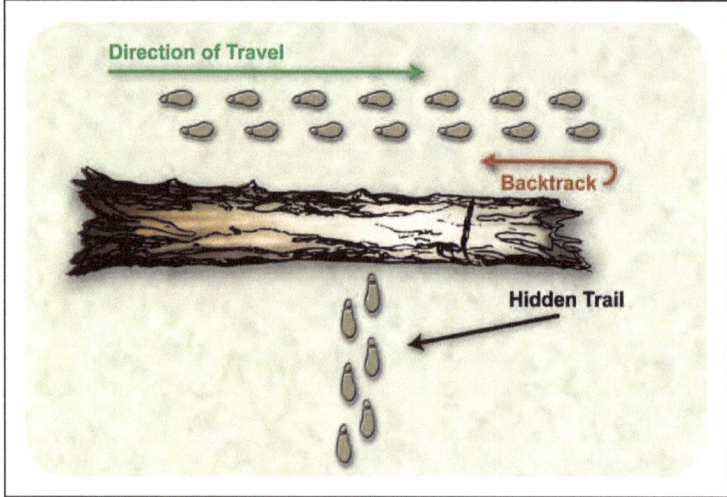

Figure 2-9. Backward-walking deception technique

Big Tree

2-12. A good deception tactic is to change directions at large trees (Figure 2-10, page 2-7). To change directions, the team moves in any given direction and walks past a large tree (12 inches wide or larger) from 5 to 10 paces, then carefully walks backward to the forward side of the tree and makes a 90-degree change in their direction of travel, passing the tree on its forward side. This technique uses the tree as a screen to hide the new trail from the pursuing tracker. A variation used near a clear area would be for the team to pass by the side of the tree that it uses to change direction on its next leg. They walk past the tree into a clear area for 75 to 100 meters and then walk backward to the tree. The team then moves 90 degrees and passes on the side away from the tracker. This method could cause the tracker to follow the team's sign into the open area where, after losing the track, he might cast in the wrong direction for the track. Generally, this technique only works on combat trackers not professional trackers.

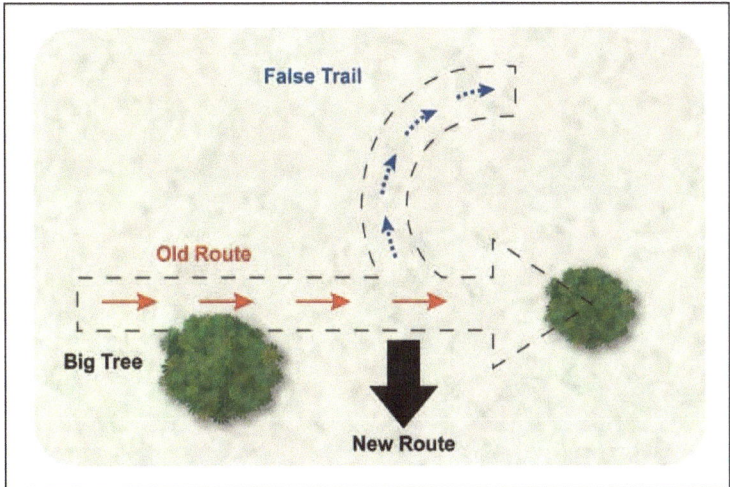

Figure 2-10. Big tree deception technique

Note. By studying signs, an observant tracker can determine if an attempt is being made to confuse him. If the team tries to lose the tracker by walking backward, footprints will be deepened at the toe and soil will be scuffed or dragged in the direction of movement. By following carefully, the tracker can normally find a turnaround point.

Cut the Corner

2-13. The team uses this deception method when approaching a known road or trail. About 100 meters from the road, the team changes its direction of movement, either 45 degrees left or right. Once the team reaches the road, they leave a visible trail in the direction of the deception for a short distance down the road. The tracker should believe the team *cut the corner* to save time. The team backtracks on the trail to the point where it entered the road and then carefully moves down the road without leaving a good trail. Once the team achieves the desired distance, they change direction and continue movement (Figure 2-11, page 2-8). Combining the big tree method with this method improves the effectiveness of this deception.

Slip the Stream

2-14. The team uses this deception method when approaching a known stream. It executes this method just like the cut-the-corner maneuver. The team establishes a 45-degree deception maneuver upstream, and then enters the stream. The team moves upstream and establishes false trails, if time permits. By moving upstream, floating debris and silt will flow downstream covering the true direction and exit point. The team then moves downstream to escape, since creeks and streams gain tributaries that offer more escape alternatives (Figure 2-12, page 2-9). To cause further confusion, the team uses false exit points. However, the team must be careful not to cause a false exit to give away its intended travel direction.

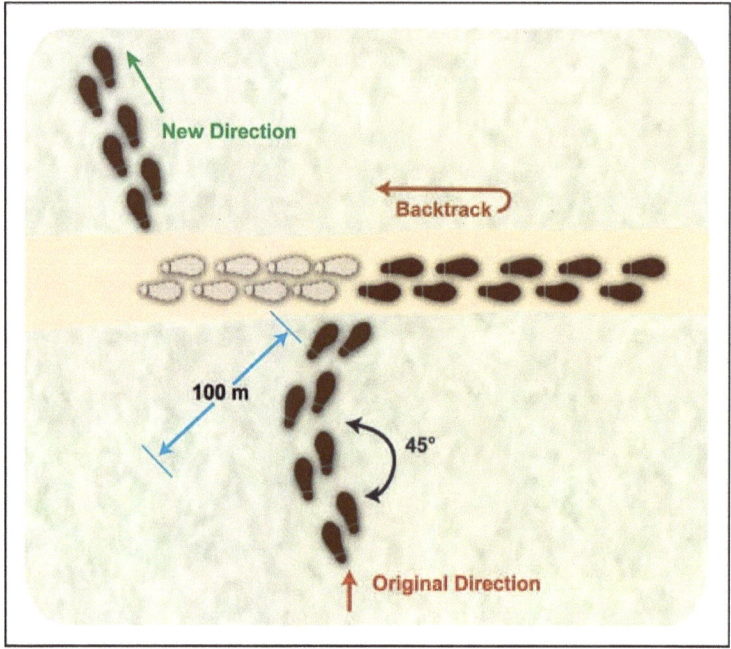

Figure 2-11. Cut-the-corner deception technique

Arctic Circle

2-15. The team should use this deception in snow-covered terrain to escape pursuers or to hide a patrol base. It establishes a trail in a circle as large as possible (Figure 2-13, page 2-9). The trail that starts on a road and returns to the same start point is effective. At some point along the circular trail, the team should remove snowshoes (if used) and carefully step off the trail, leaving one set of tracks. The team can use the big tree maneuver to screen the trail. From its hiding position, the team returns over its steps and carefully fills them with snow one at a time. This technique is especially effective if it is snowing.

Fishhook

2-16. The team uses this technique to double back on its own trail in an over-watch position (Figure 2-14, page 2-10). It can observe the back trail for trackers and prepare to ambush pursuers. If the pursuing force is too large to be destroyed, the team should strive to eliminate the tracker. It uses hit-and-run tactics, and then moves to another ambush position. The terrain must be used to the team's advantage.

2-17. Dog and visual trackers are not infallible; they can be confused with simple techniques and clear thinking. The team should not panic and try to outrun a dog or visual tracker. It only makes it easier for the tracking team. The successful tracker keeps his head and always plans two steps ahead. Even if trackers are not in the area, it is always best to use countertracking techniques.

Note. Teams must always remember there is no way to hide a trail from a professional tracker.

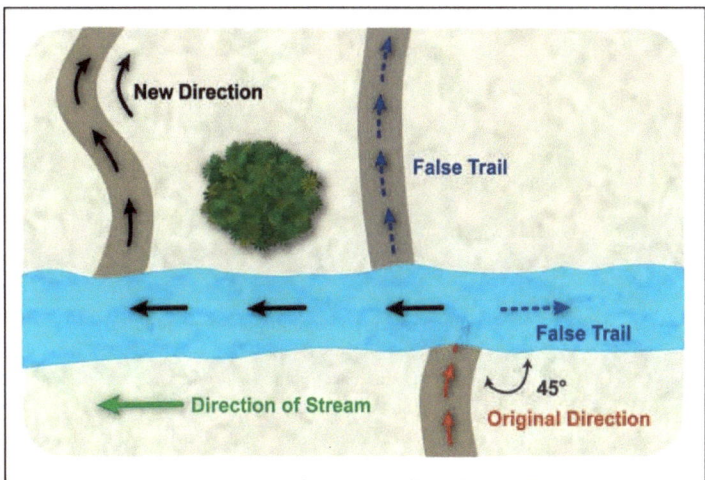

Figure 2-12. Slip-the-stream deception technique

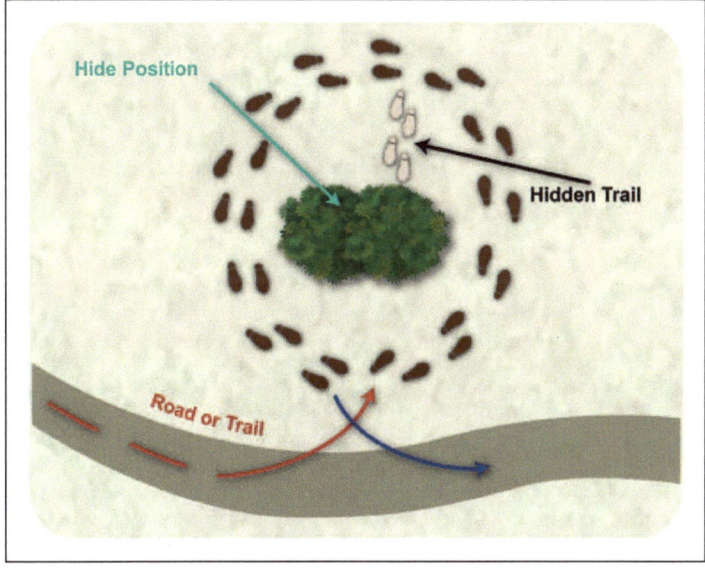

Figure 2-13. Arctic Circle deception technique

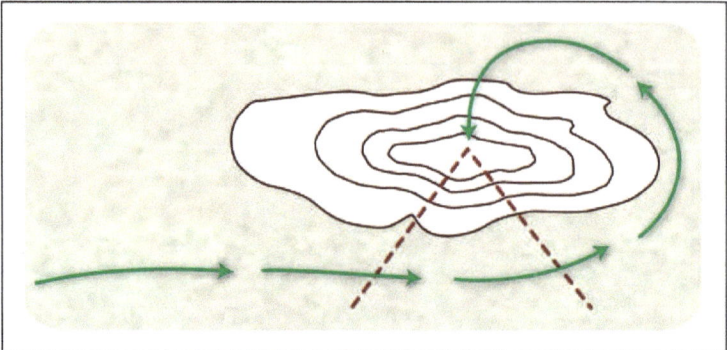

Figure 2-14. Fishhook deception technique (most popular)

Chapter 3

Dog-Tracker Teams

INTRODUCTION

Note. Throughout this chapter the term "subject" is used to refer to the entity being tracked.

3-1. When mentioning urban tracking, dog-tracker teams are the traditional tool that comes to mind. However, with modern electronic devices, a dog is only one tool capable of tracking a person's whereabouts. Traditional tracking techniques used in rural environments can also be used in urban terrain, if the tracker is proficient enough to read the signs.

3-2. The three types of tracker dogs are as follows:

- *Visual dogs* rely upon their acute vision. They usually are the final part of tracking before shifting over to the attack mode.
- *Search dogs* run free and search using airborne scents.
- *Tracker dogs* run on leashes and use ground scents.

3-3. Many myths surround the abilities and limitations of canine trackers. The first and perhaps greatest myth is that tracking involves only the dog's sense of smell. Canine tracking involves a team—a merging of man and dog. Dogs use their eyes, noses, and ears; trackers use their eyes and knowledge of the quarry. Together, they create an effective team that maximizes their strengths and minimizes their weaknesses. Another common myth is that dog teams cannot track at night; this is false, as the probability of detection is much greater at night. Handler safety in difficult terrain is the only reason to hold back search dog teams at night on a high priority search. Besides night, the next best times to track using search dog teams is early morning, late afternoon, and evening.

3-4. Midday convection currents in summer decrease a dog's effectiveness. The subject is not only trying to evade and outwit "just" a dog but also the dog's handler. The most common breeds of dog used for tracking are the shepherd (German and Dutch) and the Belgian Malinois. These dogs are trained to respond independently to a variety of situations and threats. Good tracking dogs are a rare and difficult to replace asset.

3-5. If the dog losses the trail, a visual tracker assists the dog handlers in finding a track. He can radio ahead to another tracker and give him an oral account of the track picture. A visual tracker is slower than dogs because he must always use his powers of observation, which creates fatigue. His effectiveness is limited at night.

3-6. Tracker dogs smell microbes in the earth that are released from disturbed soil. The trail has no innate scent of specific quarry, although trails do vary depending on the size and number of the quarry. For example, a scent is like the wake a ship leaves in the ocean, but no part of the ship is left in the wake. The white, foamy, disturbed water is the trail. The result is entirely different from a point smell of the quarry, such as sweat, urine, or cigarette smoke. The same training that makes tracking dogs adept at tracking a scent trail applies to finding a point smell.

3-7. Smelling is a highly complex process and many variables affect it. The most important element in tracking is the actual ground, such as earth and grass. The ground contains living microbes that are always disturbed by the quarry's passage. Artificial surfaces (concrete and macadam) and mainly inorganic surfaces (stone) provide little or no living microbes to form a scent track.

3-8. The following are some things to consider when working with or against dog-tracker teams:

- Human bodies (alive or dead) constantly shed cells and give off gases and vapors.
- The cells humans shed are lighter than air (.014 microns or smaller) and stay suspended in the air.
- Cells and odors act like smoke.
- When the sun is overhead on a calm day, smoke and scent rise up from convective currents. This is the toughest time for dogs to follow a track. The wind shears the convective column and causes the particles to follow along the ground, allowing dogs to overcome the problem.
- Days with low- or mid-level clouds reduce convection, enhancing the dog's ability to detect the scent trail.
- When shadows are longer (the sun is not overhead)—such as during the morning, evening, and in the winter—conditions are better for dogs.
- The best conditions for dogs are usually at night due to the absence of convection.
- The warmth of live bodies will cause some convective lift on cold, dead calm nights. This causes a problem on flat terrain. On hills, there is usually a downslope laminar flow of air to overcome the problem.

3-9. Cells or rafts (commonly categorized as smells) are carried in what are known as scent plumes. Scent plumes, like smoke, fall into several patterns depending on the weather. These patterns are called "fumigating," "lofting," "fanning plumes," "coning plumes," and "looping plumes."

3-10. **Fumigating** occurs when a combination of stable air aloft and unstable air at the surface meet. As the morning sun hits the surface, it rapidly warms. The cooler scent plume will then diffuse down through the warmer air and bring scents down into valleys and low spots. A dog below can easily detect a subject on a hillside. Dogs should be in the field before dawn.

3-11. **Lofting** is the reverse of fumigating. Lofting results when stable air is at the surface with unstable air aloft. Lofting occurs after the dusk and the ground is cooling but the air aloft is still warm. This is typical of valleys in the late afternoon and elsewhere in the early evening. On calm evenings where this situation occurs, handlers should work their dogs along ridges and higher slopes.

3-12. **Fanning plumes** occur at night in stable air. If the point source is on the flat, the scent will hold at the same elevation level. If the point source is on a hill, the scent could be overhead. A dog may alert on a subject across a canyon, at the same elevation, but has no way to follow them. Handlers should note and report alerts. A series of night-time alerts at the same elevation is an important clue and should initiate checks elsewhere in the area at that elevation.

3-13. **Coning plumes** are typical on cloud-covered days. The presence of clouds creates the best tracking environment for air-scenting dogs.

3-14. **Looping plumes** are typical of clear or high cloudy days and midday, high-convection situations. Scent will rise up, cool, loop back down, heat up again and, rise back up in a cycle. The dog will alert, put its head up, and then lose the scent. An experienced handler will mark the map and possibly can get a direction from a line of these alerts. Sometimes, several dogs in the field will establish the line over half a mile or so in this way, ultimately pointing to the origin of the scent.

3-15. A search or a scent-discrimination dog builds a scent picture of the person that it is tracking. Scent may be short-lived and its lifespan is dependent upon the weather and the area that the person last passed through. The sun and the wind, as well as time, destroy the scent. There are both airborne and ground scents. Airborne scents can be blown away within minutes or a few hours. Ground scents can last longer than 48 hours under ideal conditions. Bloodhounds have been known to successfully track a scent that was left behind 7 days before.

3-16. Wind and moisture are other variables that affect tracking. Foggy and drizzly weather that keeps the ground moist is best. Too much rain can wash a trail away; depending on the strength of the trail, it takes persistent, hard rain to erase a scent trail. Usually, the scent is not washed away but only sealed beneath a layer of ground water. A short, violent rainfall could deposit enough water to seal the scent track, but after the rain stops and the water layer evaporates the microbe trail would again be detectable by dogs. Hard, dry ground releases the fewest microbes and is the most difficult terrain for dogs to track on.

3-17. A dog may also have difficulty following a trail on a beach or dusty path, but his human tracker could easily follow the footprints visually. Soldiers must always remember a man and a dog are tracking them. Tracker dogs track on the trail of the subject, whereas search dogs track downwind of the trail.

3-18. Wind strength and direction are important factors in tracking. Strong wind inhibits tracking a scent trail but makes it easier for a dog to find a point scent source—like a hide. A general rule is that a dog can smell a man-sized source downwind out to 50 meters and a group-sized source—a hide—out to 200 meters under ideal conditions. Upwind, a source 1 meter away could be missed (Figure 3-1).

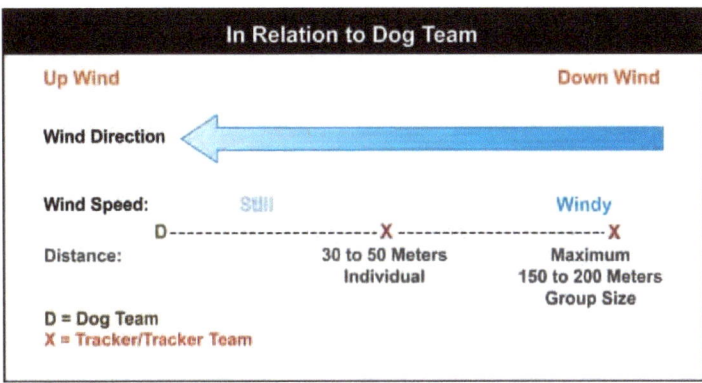

Figure 3-1. Wind strength and direction

3-19. A strong wind disperses microbes that arise from the ground, hindering a dog's ability to follow a trail. However, a strong wind increases the size of a point scent, helping a dog to find the target in an area search.

3-20. An inflexible rule for the life of a scent trail cannot be provided. In Germany, trackers rate their chances of following trails more than three days old as negligible. Terrain, weather, and the sensitivity of the dog are some of the many variables that affect the scent trail. A point smell will last as long as the target emits odors.

3-21. Although dogs are mainly scent hunters, they also have good short-range vision. Dogs are colorblind and do not have good distance vision (camouflage works extremely well against dogs). However, they can detect slight movements. Dogs also have a phenomenal sense of hearing, extending far beyond human norms in both the frequency range and in sensitivity. Dogs use smell to approximate a target and then rely on sound and movement to pinpoint that target.

3-22. Dogs have tremendous detection abilities; however, they also have limitations. Following a scent trail is the most difficult task a dog can perform. The level of effort is so intense that most dogs cannot work longer than 20 to 30 minutes at a time, followed by a 10- to 20-minute rest. Dogs can perform this cycle no more than 5 or 6 times in a 24-hour period before reaching complete exhaustion. The efficiency of the search also decreases as the dog tires. In wartime, the situation forces the maximum from men and equipment, but times should remain constant for dogs because they always give 100 percent. If the subject keeps moving and stays out of the detection range of the dog and handler, the subject could outlast scent trackers.

3-23. When looking for subjects, trackers mainly use wood line sweeps and area searches. A wood line sweep consists of walking the dog upwind of a suspected wood line or brush line, the key is upwind. If the wind is blowing through the woods and out of the wood line, trackers move 50 to 100 meters inside a wooded area to sweep the wood's edge. Since wood line sweeps tend to be less specific, trackers perform them faster. Trackers perform an area search when a subject's location is specific, such as a small wooded area or block of houses. If possible, the search area is cordoned off and the dog-tracker teams are brought on-line about 25 to 150 meters apart, depending on terrain and visibility. The handlers then advance, each moving their dogs through a specific corridor. The handler controls the dog entirely with voice commands and gestures. He remains undercover, directing the dog in a search pattern or to a likely target area. The search line moves forward with each dog dashing back and forth in assigned sectors.

3-24. A tracker should take every opportunity to practice applying countertracking techniques in all types of terrain and weather conditions. When planning training opportunities, the tracker should consider the following:

- *Wind.* A moderate breeze of 13 to 18 miles per hour (mph) is needed to overcome high convection. If dust and small branches are moving, the wind is 13 mph or more.
- *Eddies.* They can form at bends in canyons and at the mouth of tributaries, bringing scents from different directions. Eddies can also form at the edges of meadows, behind hedgerows, and at any break in vegetation. The dog-tracker team should check all edges because the breeze may not have carried the scent away.
- *Large Roll Eddies.* They can form on the lee of ridges and canyon rims and can cause upslope winds that blow opposite of prevailing winds.
- *Ridge Top Saddles and Mountain Passes.* These increase wind flow and are good places to pick up air scents.
- *Forest Openings.* Openings in a forest will heat up and bring in drafts from all directions. The dog-tracker team should check the middle of the opening to take advantage of this.
- *Single Trees and Telephone Poles.* In a field a single tree or telephone pole can act like a chimney and create a vortex that attracts rafts of scent. This vortex can provide an opportunity similar to the forest opening described above.
- *Breeze.* A 20-mph fresh breeze will usually be slowed down to 4 mph or less in a dense forest. In contrast, a 4-mph breeze will generally only be slowed to approximately 2.5 mph.
- *Updrafts.* When the sun is shining on the slopes of a hill, there will normally be updraft airflow. Major canyons will normally have an upstream breeze during the day. This will cause the updrafts on the slopes to move diagonally upslope and upstream. Updrafts increase in velocity as they rise, which causes ridges to receive scents from the entire slope.
- *Downdrafts.* When the side slopes go into shadow, a downdraft begins. During these conditions it is best to search from the bottom up. Downdrafts flow down like water and act like a dam around debris pile hollows, low shady spots, and brush for scent pooling. A good place to check in the shade and at night is the mouths of side drainages from the terrain creating the downdraft.
- *Thunderstorms.* Thunderstorms create downdrafts that push air out in all directions from directly under the cell at the mature stage. A dog can alert from a great distance, so the Soldier should note the location of the thunderhead and the wind direction when the dog alerts. The thunderhead will have a strong convective updraft and will suck air toward it before reaching the mature stage when it starts to rain. It is important to pay attention to what is happening and note the time and conditions when the dog alerts since the wind may change directions and speed rapidly.
- *Smoke Candles.* These are a good tool for the tracker to use during practice sessions. They help him visualize the effects of air flow.

TECHNIQUES TO DEFEAT DOG-TRACKER TEAMS

3-25. Although dog-tracker teams are a potent threat, there are counteractions available to the subject. As always, the best defenses are basic infantry techniques: good camouflage and light, noise, and trash discipline. Dogs find a subject either by detecting a trail or by a point source, such as human waste odors at the hide site.

3-26. It is critical to try to obscure or limit trails around the hide, especially along the wood line or area closest to the target area. Surveillance targets are usually major axes of advance. Trawling the wood lines along likely-looking roads or intersections is a favorite tactic of dog-tracker teams. When moving into a target area, the subject should take the following countermeasures:

- Remain as far away from the target area as the situation allows.
- Never establish a position at the edge of cover and concealment nearest the target area.
- Minimize the track. Try to approach the position area on hard, dry ground or along a stream or river.
- Urinate in a hole and cover it up. Never urinate more than once in exactly the same spot.
- Deeply bury fecal matter. If the duration of the mission permits, the subject can use meals, ready to eat bags, sealed with tape and take them with him when leaving. (If using this technique, he ensures the bags are sealed air tight to avoid odors escaping.)
- Never smoke.
- Carry all trash until it can be buried elsewhere.

3-27. Dogs tracking a subject use odors left behind or around the subject to locate him. Sweat from exertion or fear is one of these. Wet clothing or material from damp environments holds in the scent. Soap or deodorant used before infiltration helps the dogs to find the subject. Foreign odors, such as oils, preservatives, polish, and petroleum products aid the dogs. If time permits, the subject should try to change his diet to that of the local inhabitants before infiltration.

3-28. When the subject first arrives in the operational area, it is best for him to move initially in a direction that is 90 to 170 degrees away from his objective. Objects or items of clothing not belonging to the subject should be carried into the operational area in a plastic bag. When the subject first starts moving, he should drop an item of clothing or piece of cloth out of the bag and leave it on a back trail. This step can confuse a dog long enough to give the subject more of a head start. If dogs are brought in later, the subject's scent will be very faint; whereas, the scent from the dropped item will still be strong.

3-29. While traveling, the subject should try to avoid heavily foliaged areas, as these areas hold the scent longer. When the situation permits, the subject should periodically move across an open area that the sun shines on during the day and has the potential of being windswept. The wind moves the scent and will eventually blow it away; the sun destroys scent very rapidly.

3-30. When the situation permits, the subject should make changes in direction at the open points of terrain to force the dog to cast for a scent. If dogs are very close behind, moving through water does not confuse them, as scent will be hanging in the air above the water. Moving through water will only slow the subject down. Throwing CS gas to the rear or using blood, spice mixtures, or other concoctions will prevent most dogs from smelling the subject's scent, but it is not effective on a trained tracker dog.

3-31. Although water will not confuse a dog if he is close, running water, such as a rapidly moving stream, will confuse a dog if he is several hours behind. However, areas with foliage, stagnant air, and little sunlight will hold the scent longer. Therefore, the subject should try to avoid any swampy areas.

3-32. To avoid confusing the scent picture of the dog, the dog-tracker team should not move through areas frequently traveled by other people. On the other hand when more than one subject is being tracked, the subjects should split up from time to time to confuse the dogs by dissipating the scent. The best place to split up is in areas frequently traveled by indigenous personnel.

3-33. If a dog-tracker team is on the subject's trail, the subject should not run because their scent will become stronger. The subject may attempt to wear out the dog handler and confuse the dog, but should always be on the lookout for a good ambush site to fishhook into. If it becomes necessary to ambush the tracking party, the subject should fishhook into the ambush site and kill or wound the handler, not the dog. A tracker dog is trained with his handler and will protect him should the handler become wounded. This practice allows the subject to move off and away from the area while the rest of the tracking party tries to give assistance to the handler.

3-34. If a dog-tracker team moves into the area, the subject should first check wind direction and strength. If the subject is downwind of the estimated search area, the chances are minimal that the subject's point smells will be detected. If upwind of the search area, the subject should attempt to move downwind. Terrain and visibility dictate whether the subject can move without being detected visually by the handlers.

Sweeps are not always conducted just outside of a wood line. Wind direction determines whether the sweep will be parallel to the outside or 50 to 100 meters inside the wood line.

3-35. The subject has options if caught inside the search area of a line search. Handlers rely on radio communications and often do not have visual contact with each other. If the subject has been localized through enemy radio detection-finding equipment, the search net will still be loose during the initial sweep. A single subject has a small chance of hiding and escaping detection in deep brush or in woodpiles. Larger groups will almost certainly be found. Yet, the subject may have the chance to eliminate the handler and to escape the search net.

3-36. If the handler feels threatened, he will hide behind cover with the dog. He searches for movement and then sends the dog out in a straight line toward the front. Usually, when the dog has moved about 50 to 75 meters, the handler calls the dog back. The handler then moves slowly forward and always from covered position to covered position. Commands are by voice and gesture with a backup whistle to signal the dog to return. If a handler is killed or badly injured after releasing the dog but before recalling it, the dog continues to randomly search out and away from the handler. The dog usually returns to another handler or to his former handler's last position within several minutes. This time lapse creates a gap from 25 to 150 meters wide in search pattern. Response times by the other searchers tend to be fast. Given the high degree of radio "chatter," the injured handler will probably be quickly missed from the radio net. Killing the dog before the handler will probably delay discovery only by moments. Dogs are so reliable that if the dog does not return immediately, the handler knows something is wrong.

3-37. If the subject does not have firearms, he must realize human versus dog combat is hazardous. If a knife or large club is available, it is possible for a subject to deal with one dog with relative ease. The subject must keep low and use his wrist to strike upward, never overhanded. Dogs are quick and will try to strike the groin or legs. Most attack dogs are trained to go for the groin or throat. Tracking dogs are more likely to attack the arms, legs, or groin. Regardless the type of dog, before attempting to defeat the dog by stabbing it, the subject should wrap his arms and legs with as much clothing as possible to reduce the bite effects of the dog. If alone and faced with two or more dogs, the subject should flee the situation.

3-38. Small, lightly-armed dog-tracker teams are a potent threat to a subject because dogs greatly increase the area that a rear area security unit can search. Due to the dog-tracker team's effectiveness and its lack of firepower, a subject may be tempted to destroy such an "easy" target. Whether a subject should fight or run depends on the situation.

3-39. Eliminating or injuring the dog-tracker team only confirms to security forces that there is a hostile subject operating in the area. The techniques for attacking a dog-tracker team should be used only in extreme situations or as a last measure. The preferred method of defeating a dog team is to exhaust the handler, thus rendering the dog team useless while not providing solid evidence of the subject's presence in the area.

Appendix A
Tracking Log

Soldiers should keep a log to record specific information about tracking signs they have observed, in case the source is not found on the first effort. Below is a sample tracking log (Table A-1) listing some ideas as to the minimum amount of data the Soldier should consider recording about the signs. This particular sample uses a footprint, but the tracker can tailor the log to fit the situation.

Table A-1. Sample tracking log

Date:	Time:
Location:	
Subject Heading:	
Basic Type:	
Pattern:	
Dimensions Overall:	
L: W:	
Stride (Heel-to-Heel)	
Ground:	
Remarks:	

Appendix B

Tracking Log for Training Purposes

During training exercises, Soldiers can use the sample tracking log shown below (Table B-1). The log provides the Soldier with detailed information to consider when conducting training exercises.

Table B-1. Sample tracking log for training purposes

Location:			
Start Date:_____	Day:_____	Time:_____	Mileage:_____
End Date:_____	Day:_____	Time:_____	Mileage:_____
		Total Hours:_____	Total Miles:_____
Tracker:			

Activity: Time:_____		Time:_____
_____	UTS: _____	
_____		
_____	Other: _____	
Miscellaneous:		
Explanation:		

Conditions:			
Weather:_____	Wind:_____	Humidity:_____	
Clouds (percent):_____	Daylight:_____	Darkness:_____	
Trail Age (hours):_____	Number of Subjects:_____		

Area Type:			
□ Gravel Pavement	□ Pine Needles	□ Rocky	□ Meadows
□ Sand/Loam	□ Chaparral	□ Woodlands	□ River Canyon
□ Leaf Litter	□ Open Desert	□ Grasslands	□ Urban

Comments:

Glossary

SECTION I – ACRONYMS AND ABBREVIATIONS

ARNG	Army National Guard
ARNGUS	Army National Guard of the United States
CS	2-chlorobenzalmalononitrile
mph	miles per hour
SF	Special Forces
SSE	sensitive site exploitation
TC	training circular
TTP	tactics, techniques, and procedures
USAJFKSWCS	United States Army John F. Kennedy Special Warfare Center and School
USAR	United States Army Reserve

References

REQUIRED REFERENCES
These documents must be available to intended users of this publication.

None

RELATED PUBLICATIONS
These documents contain relevant supplemental information.

ARMY PUBLICATIONS

FM 23-10, *Sniper Training,* 17 August 1994

NONMILITARY

http://www.sarbc.org/dinfo.html

REFERENCED FORMS
The following form is available on the AKO, AHP, and APD Web sites

DA Form 2028 (Recommended Changes to Publications and Blank Forms)

Index

A
artificial surfaces, 3-1,

B
bloodhounds, 3-2

C
cell, 3-2

convection,
    absence, 3-2
    reduce, 3-2

convective currents,
    column, 3-2
    lift, 3-2
    updraft, 3-4

D
deception techniques,
    Arctic Circle, 2-8 and 2-9
    big tree, 2-6 and 2-7
    cut the corner, 2-7 and 2-8
    fishhook, 2-1, 2-10, 3-5
    slip the stream, 2-7, 2-9
    walking backward, 1-5 and
       1-6, 2-1, 2-6 and 2-7

dogs,
    search, 3-1 through 3-3
    tracker, iv, 2-1, 2-10, 3-1
       through 3-6
    visual, 2-10, 3-1

F
foot wrapping, 2-2 through 2-4

fumigating, 3-2

I
immediate-use intelligence,
    1-2, 1-19

K
key prints, 1-7

L
laminar flow, 3-2

litter, 1-2, 1-18, 1-20

lofting, 3-2

M
microbes, 3-1 through 3-3

O
odors, 1-18 and 1-19, 3-2
    through 3-5

P
plumes,
    coning, 3-2
    fanning, 3-2
    looping, 3-2

point,
    smell, 3-1, 3-3, 3-6
    source, 3-2, 3-4

R
rafts, 3-2

S
scent,
    airborne, 3-2
    concoctions, 3-5
    discrimination, 3-2
    ground, 3-2
    track, 3-2

signs,
    brushed out, 2-5
    brushing out, 2-4
    ground, 1-2
    high, 1-2
    permanent , 1-2
    temporary, 1-2

source,
    group-sized, 3-3

    hide, 3-3
    man-sized, 3-3

stains, 1-13 through 1-16

T
tracker, visual, 2-10, 3-1

tracking indicators,
    box edge, 1-3
    displacement, 1-2 and 1-15
       signs, 1-8
    heel push, 1-5
    toe digs, 1-5

tracking log, A-1, B-1

tracking, urban, 3-1

trail signs, reduction of, 2-6

training considerations,
    breeze, 3-4
    eddies, 3-4
       large roll, 3-4
    downdrafts, 3-4
    forest openings, 3-4
    mountain passes, 3-4
    ridge top saddles, 3-4
    single trees, 3-4
    smoke candles, 3-4
    telephone poles, 3-4
    thunderstorms, 3-4
    updrafts, 3-4

W
weathering, 1-2, 1-16 through
    1-20

wind
    direction, 1-18, 3-3 and 3-4
    strength, 1-20, 3-3
    swept, 3-5

TC 31-34-4
30 September 2009

By Order of the Secretary of the Army:

GEORGE W. CASEY, JR.
General, United States Army
Chief of Staff

Official:

JOYCE E. MORROW
Administrative Assistant to the
Secretary of the Army
0923904

DISTRIBUTION: *Active Army, Army National Guard, and United States Army Reserve:* To be distributed in accordance with initial distribution number 116002.

www.ingramcontent.com/pod-product-compliance
Lightning Source LLC
Chambersburg PA
CBHW041205180526
45172CB00006B/1198